EARLY FLYING DESIGNS

Early AUTOGIRO
(1923)

U. S. MAIL PLANE
(1925)

Lindbergh's "THE SPIRIT OF ST. LOUIS"
First New York-to-Paris Solo Flight
(1927)

TRAVELAIR SPEED PLANE
(1930)

Von Opel's ROCKET PLANE
(1930)

The Book of
AIRPLANES

By GEORGE J. ZAFFO

ISBN: 0-448-02243-5 (Trade Edition)
ISBN: 0-448-03730-0 (Library Edition)
Library of Congress Catalog Card Number: AC 66-10901

1973 PRINTING
GROSSET & DUNLAP • Publishers • NEW YORK

JET PLANE IN FLIGHT • One of the fastest operational jet airplanes is the Convair Hustler, which weighs 165,000 pounds fully loaded and can fly at 1,385 miles per hour. It has four turbojet engines attached under its wings. The men of the Strategic Air Command call it the B-58A Hustler. This fast bomber, carrying a crew of only three men, can fly 2,500 miles without refueling at an altitude of 60,000 feet. The delta wing enables it to fly faster than planes whose wings are straight. Additional fuel, material or weapons may be contained in a pod under the fuselage. The Hustler's wing span is 56 feet 10 inches; its fuselage is 96 feet 9 inches long; its height is 31 feet 5 inches.

5671

Zaffo

AXES OF CONTROL

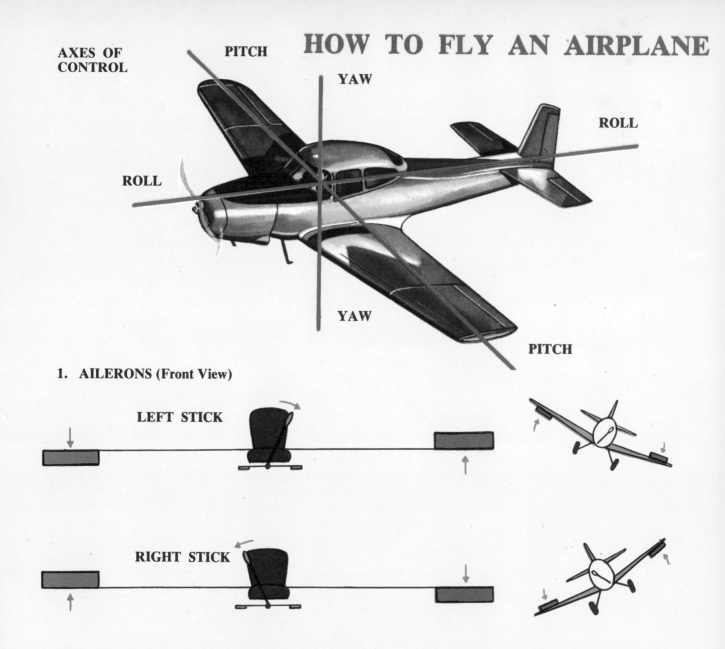

1. **AILERONS (Front View)**

LEFT STICK

RIGHT STICK

When you sit in an airplane cockpit you see a whole bank of knobs, dials and handles. They look complicated, but the controls that steer the airplane are really very simple to operate. An automobile needs only a steering wheel. It drives on the road and turns only to one side or the other. But an airplane not only turns right or left but also up and down. An airplane operates on three fixed lines, called axes—the axis of pitch, the axis of roll, and the axis of yaw. Movement around the axis of yaw is controlled by the rudder. Movement around the axis of pitch is controlled by the elevators. Movement around the axis of roll is controlled by the ailerons.

There are two ways to make an airplane turn sideways. One way is to move the *ailerons*. The ailerons are little hinged flaps on the back end of each wing. They can be moved up or down by a *control stick* in the cockpit. When the control stick is moved to the right, the aileron on the right wing moves up and the aileron on the left wing moves down. This makes the air blowing over the ailerons push the right wing down—and the airplane starts to turn or *bank* to the right. When the control stick is moved to the left, the left wing is pushed down in just the same way, and the airplane starts turning to the left. The other way to make the airplane turn sideways is to move the *rudder*. The rudder is in the back of the airplane, and it works like the tail of a fish. The rudder is operated by foot pedals. When the right pedal is pushed down, the rudder turns to the right. The wind blows against it and pushes the tail to the left. That makes the nose of the

2. RUDDER (Top View)

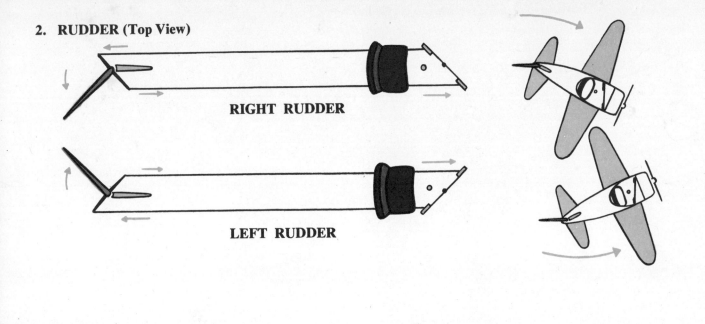

RIGHT RUDDER

LEFT RUDDER

3. ELEVATORS (Side View)

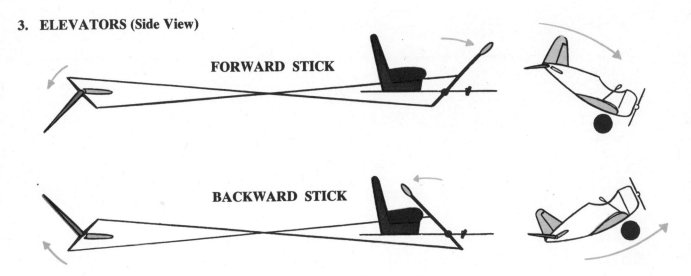

FORWARD STICK

BACKWARD STICK

plane turn to the right. The nose turns to the left when the left rudder pedal is pushed down. An airplane turns best when both the control stick and the rudder pedals are pushed at the same time, since they work together.

There are also two ways to make the plane go up and down. One way is to move the *elevator*. The elevator is a little hinged flap set crosswise in the tail. The same control stick which moves the ailerons also operates the elevator. When the control stick is pushed forward, the elevator moves down. The air blows against it and pushes the tail up. This makes the nose of the airplane point down. Then the plane starts to go down. When the control stick is pulled backward, the tail is pushed down, and the nose of the plane points up. Then the plane starts to climb. The other way to make the plane go up or down is to make the engine and propeller turn faster. The engine is controlled by the *throttle*—just as in an automobile. When the throttle is pushed in, the engine turns faster. Then the airplane starts to climb. When the throttle is pulled out, the engine runs slower and the airplane starts to descend. The throttle is also used to make the airplane go faster or slower, of course. So you see a plane can be steered through the air with only three controls. They are the control stick, used much like the steering wheel of a car, the rudder pedals, and the throttle. It is a little harder to fly an airplane than to drive an automobile, but not much harder. Maybe you'll be a pilot someday!

TWIN CO-AXIAL ROTORS

TWIN INTERMESHED ROTORS

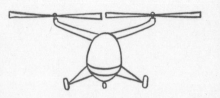

TWIN OUTBOARD ROTORS

TWIN TANDEM ROTORS

DIFFERENT WAYS TO MOUNT TWIN ROTORS

THE HELICOPTER • Helicopters cannot fly as fast as airplanes, but they can do some things that airplanes cannot do: they can land straight down or take off straight up. Helicopters do not have wings. Instead, they have two sets of propellers, called *rotors,* that whirl around very fast and lift the machines right off the ground. Most helicopters have one big rotor overhead and a small rotor in the tail. The tail rotor helps the pilot steer the helicopter and keep it going in the right direction. If there were no tail rotor, the helicopter would spin around opposite to the movement of its overhead rotor. The big overhead rotor lifts the helicopter straight up or

**SINGLE ROTOR
AND TAIL ROTOR**

**SINGLE ROTOR
JET REACTION TAIL**

**SINGLE ROTOR
JET-PROPELLED BLADES**

**SINGLE ROTOR
PROPELLERS ON BLADES**

**DIFFERENT WAYS TO
MOUNT SINGLE ROTORS**

lets it come down straight. It also helps it move ahead through the air, because when the rotor blade is turning, it takes a bigger "bite" through the air on its way back than it does when moving forward. (A canoe paddle in water works the same way.) Some helicopters have two big overhead rotors and no tail rotors. Each of the big rotors helps lift the helicopter and move it forward. Such helicopters do not need tail rotors to keep them from spinning, because the overhead rotors turn in opposite directions and balance each other. The picture shows helicopters picking up supplies from an aircraft carrier for combat troops on land.

PARTS OF AN AIRPLANE

1. Ailerons
2. Trailing Edge
3. Fuselage
4. Elevators
5. Rudder
6. Fin
7. Wing
8. Leading Edge
9. Windshield
10. Jet Pod
11. Horizontal Stabilizer
12. Landing Gear
13. Tail Light
14. Wing-Tip Fuel Tank
15. Air Intake
16. Jet Thrust
17. Wing Span
18. Cabin
19. Landing Light
20. Cabin Door

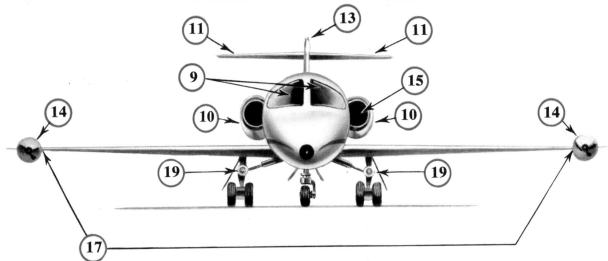

AERONAUTICAL TERMS

AILERON. A section of the rear edge of the airplane wing, which has hinges to help make the airplane turn.

AIRPLANE. An aircraft with fixed wings, which is heavier than the air and is driven by an engine.

AIRPORT. A place where airplanes land and take off. It usually has hangars where airplanes can be housed, and buildings for passengers and freight.

AIRSHIP. A lighter-than-air ship which is driven by motors and propellers.

ALTIMETER. An instrument that measures how high an aircraft is above sea level at any time.

AMPHIBIAN. An airplane that can land and take off from both land and water.

AUTOGIRO. An aircraft in which rotors, big overhead propellers, are used instead of a wing, to keep the plane aloft.

AUTOPILOT. A device that flies the airplane automatically.

BALLOON. A lighter-than-air ship which has no engine to drive it.

BIPLANE. An airplane with both a top and bottom wing.

BLIMP. An airship with no rigid frame.

BOMBARDIER. The member of a military bomber crew who aims and releases the bombs.

CABIN. The enclosed part of the airplane, that carries cargo, passengers, and crew.

COCKPIT. The part of the airplane that is used for passengers and pilots. If the space is enclosed, it is usually called a cabin instead of a cockpit.

COWLING. A removable covering over some part of the airplane, usually the engine or cockpit.

ELEVATOR. A section of the airplane tail, which moves on hinges to make the plane go up or down.

ENGINE. The machine that drives the airplane's propellers or jet stream.

FLAP. A movable hinged portion of the rear of the wing, that enables the airplane to climb and fly fast or slow.

FLYING BOAT. A kind of seaplane in which the body itself serves as a boat for landing on water.

FUSELAGE. The body of the airplane, to which the wings and tail are attached.

GLIDER. A kind of airplane which soars through the air without an engine.

HELICOPTER. A kind of aircraft with big overhead propellers, called rotors, which lift the craft straight up, and also move it forward or backward.

INSTRUMENT PANEL. A board on which the plane's instruments are installed.

LANDING GEAR. The wheels and supporting structure on which the airplane lands.

MONOPLANE. An airplane with only one wing.

NACELLE. An outside structure which encloses part of the airplane, such as the engine.

ORNITHOPTER. A kind of airplane that flies by flapping its wings.

PARACHUTE. A kind of big umbrella used to slow the fall of a man or of cargo through the air.

PILOT. The man who flies an aircraft. He is also called an aviator.

PITOT TUBE. A tube with an open end which helps measure how fast a plane is going.

POD. Attached under the wing or fuselage of an airplane. It can contain a jet engine, fuel or nuclear weapon(s).

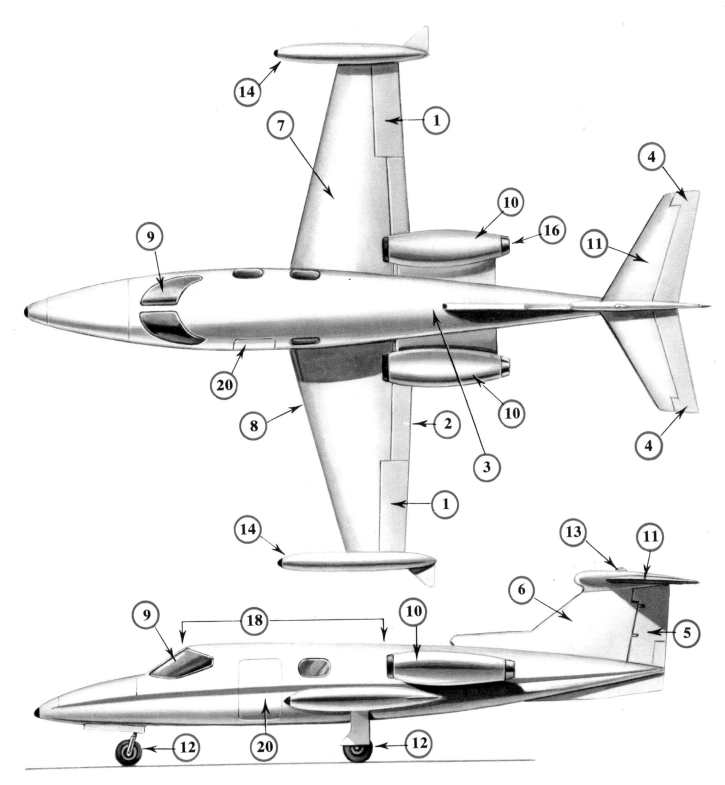

PROPELLER. A device used to push or pull a plane through the air.

RIP CORD. The rope used to open a parachute.

RUDDER. A hinged portion of the tail which helps steer the airplane from one side to the other.

SAILPLANE. A fast kind of glider.

SEAPLANE. An airplane that will land on or take off from water.

SKI. A landing gear used to land planes on snow.

SOAR. To fly without power.

STRUT. An outside support to strengthen the wing.

TABS. Hinged pieces at ends of rudder, elevator, and ailerons, which help balance the plane.

TAXI. To operate an airplane on the ground or on water.

THROTTLE. A control which regulates the speed of the engine.

TRAILING EDGE. The rear edge of a wing or propeller.

TRANSMITTER. A radio set that sends messages.

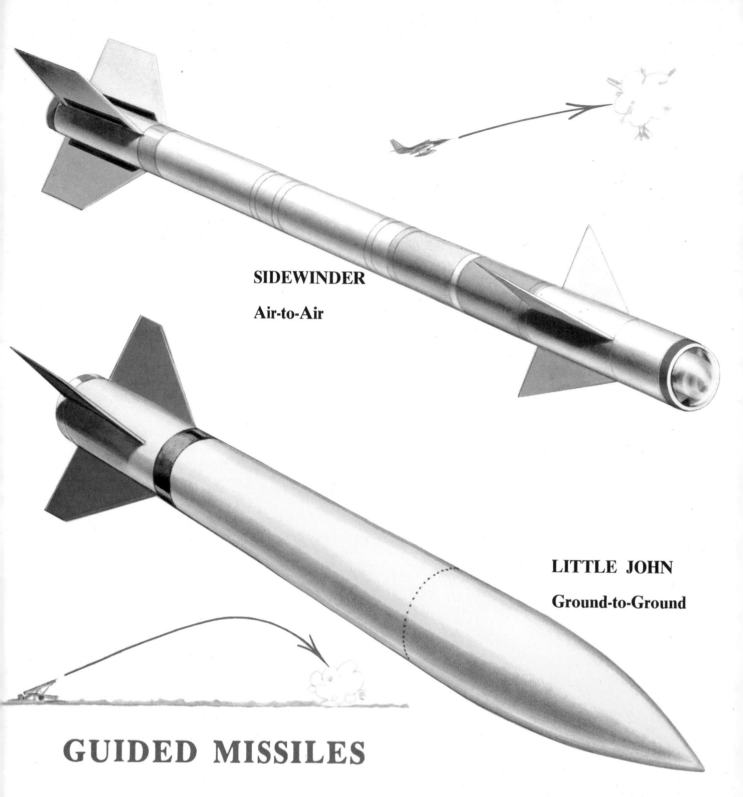

SIDEWINDER

Air-to-Air

LITTLE JOHN

Ground-to-Ground

GUIDED MISSILES

Guided missiles are airplanes or rockets which have engines to drive them, but are flown by mechanical pilots. They can be powered by liquid-fueled rocket engines, by jet engines, or by engines driving propellers. But they all have one thing in common. They are not just *aimed*, like an artillery shell or a rifle bullet. They are able to *hunt* their own targets and hit them, or else they are guided by radio or television to their targets while they are in flight. There are four types of guided missiles:

1. AIR-TO-AIR MISSILES • These are launched from airplanes and aimed at other airplanes or missiles. They can fly very fast but only for a few miles. The best air-to-air missiles are able to follow a target and hit it no matter how hard it tries to get away.

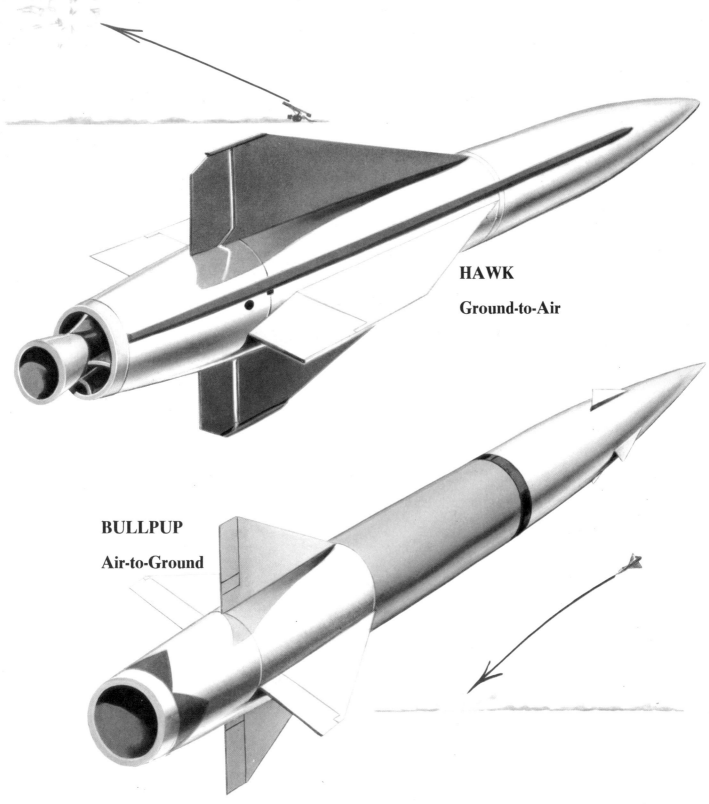

HAWK

Ground-to-Air

BULLPUP

Air-to-Ground

2. GROUND-TO-GROUND MISSILES • These are usually long-range missiles, like the German V-2 rocket. Some day, scientists believe, they may be able to hit targets 10,000 miles away. Really successful ones will not be developed for many years, however.

3. GROUND-TO-AIR MISSILES • A kind of antiaircraft weapon, they are fired from the ground to destroy enemy airplanes or enemy-guided missiles. They can be guided by radio or radar and propelled by rockets or jet engines. But they must be very fast.

4. AIR-TO-GROUND MISSILES • These are really guided bombs. Some of them automatically seek out their targets and hit them. For instance, they may have a kind of radar that is able to "smell" a steel ship or a hot factory smokestack, and fly right toward it. Such missiles may have wings and may have a rocket in the tail to make them fly faster and farther.

PAN-AMERICAN'S JET CLIPPER, BOEING 707-321B • A growing popularity and need for fast, dependable travel across oceans gave rise to the development of this great new airliner, which can carry a full load of 174 passengers between capital cities of Europe and the United States. Its range is 6,000 miles — enough to fly nonstop from Chicago to Amsterdam, Seattle to Tokyo, or New York to Rio de Janeiro. The Jet Clipper is tremendous — it has a tail fin as high as a three-story building! The wing

span alone is 145 feet 8 inches, a distance longer than that of the first flight in an airplane by the Wright Brothers. Four jet engines produce three to four times the power used to run the average ocean-going freighter. When fully loaded, the Jet Clipper weighs 325,000 pounds, yet it can fly safely on three engines and stay in the air with two engines cut off. Its cruising speed at 42,000 feet is 575 miles per hour. The Jet Clippers are flown to all six continents and on round-the-world service.

HOW ENGINES OPERATE

RECIPROCATING ENGINE

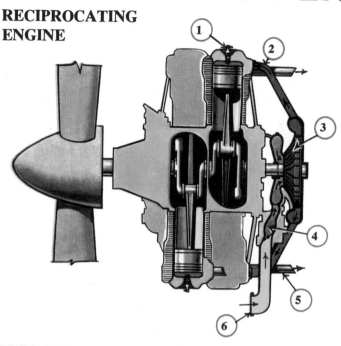

TURBOJET

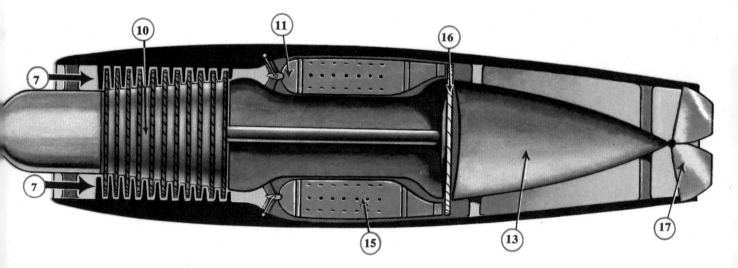

KEY TO NUMBERS

1. Spark Plug
2. Cylinder Inlet
3. Shaft-Driven Supercharger
4. Carburetor
5. Cylinder Exhaust
6. Air Scoop
7. Air Inlet
8. Propeller
9. Reduction Gear
10. Compressor
11. Fuel Spray
12. Compressor-Drive Turbine
13. Tail Cone
14. Propeller-Drive Turbine
15. Combuster
16. Turbine
17. Adjustable Exhaust Nozzle
18. Fuel
19. Oxidant
20. Pump Unit
21. Combustion Chamber
22. Exhaust Nozzle
23. Shock Wave
24. Fuel Injection
25. Flame Holder

The engine that turns a propeller is called a *reciprocating* engine. Air comes into the engine through an air scoop on the outside of the plane. Then the air goes into a supercharger, where it is forced into a very small space. After that, the air is forced into a cylinder and compressed into a space even smaller. The fuel is burned in the cylinder. When the fuel burns, it causes the air to expand and take up more space. As the air expands, it pushes a piston. Then the air escapes through the cylinder exhaust. Every time a piston is pushed by the expanding air, the piston helps turn a crankshaft. And on the end of the crankshaft is the propeller, which turns rapidly. The propeller thrusts the air behind it, and causes the plane to move through the air.

Air blown into a balloon is compressed by forcing it into a smaller space inside the balloon. Then, by releasing the balloon, air spurts out of its mouth, causing the balloon to move in the opposite direction. That's how a jet engine works. In a *turbojet,* air flows into an inlet and is forced into a small space by a compressor. Fuel is sprayed and then burned in the combuster. The heat

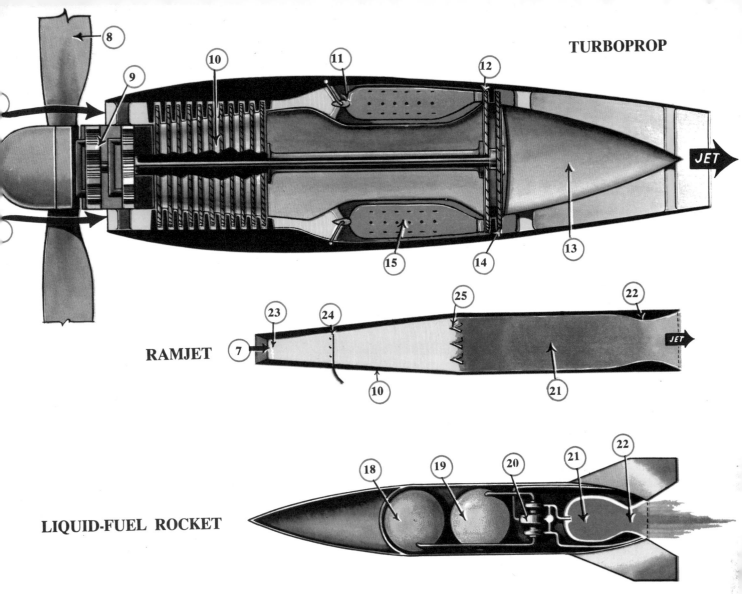

TURBOPROP

RAMJET

LIQUID-FUEL ROCKET

causes the air to expand and spurt out of an exhaust nozzle that may be adjusted to control the thrust. Just before the air escapes, it turns a turbine wheel which causes the compressor to operate, so it does an extra job, besides causing the airplane to move forward.

The *turboprop* engine has a turbine wheel, too, but in addition to operating the compressor, it is attached to a crankshaft, which turns a propeller.

One of the simplest airplane engines is the *ramjet*. It looks like a simple tube or pipe. In fact, some people call it a flying stovepipe. The air intake is narrow, shaped so that the air coming into the engine has to slow down. The air is thus compressed and forced into a small space. Then fuel is injected and burned. As the fuel burns, it heats the air, which expands, rushing out of the exhaust nozzle at a much faster speed than when it entered the air intake. The ramjet is best for fast planes, but it is also being used as a helicopter engine, with one ramjet on the tip of each rotor.

The fastest airplane engines are called rockets. They are good engines for flying at great heights and for flying very fast. But they can fly for only a short time because they burn so much fuel. A rocket engine carries its own fuel, plus oxygen needed to burn the fuel. One kind of rocket is called a *liquid-fuel rocket*. It has two tanks — one for fuel, the other for the "oxidant" which when mixed with the fuel enables it to burn without outside air. The engine usually operates with a pump unit. Very little of the liquid fuel is used to operate the pump, which mixes the oxidant and fuel. The mixture goes into the combustion chamber, where it burns at a very hot temperature, and escapes through the exhaust nozzle.

A milestone in aviation history was reached in the fall of 1955, when Pan American World Airways placed orders for the first fleet of jet airliners—45 in all—to usher in the jet air age.

Jet clippers fly high above clouds and weather, in the deep purple-blue of the substratosphere, where passengers can actually see the stars at noon! Transatlantic crossings from New York to Paris take only 6 hours 35 minutes. The longest nonstop flight Pan American schedules is from New York to Buenos Aires, a distance of 5,310 miles, lasting ten and a half hours — longer than a flight from New York to Moscow.

The day seats in the cabin section of a Jet Clipper convert into comfortable beds at night, with both upper and lower berths as large as standard single beds in your home. Imagine falling asleep six or seven miles above the world, with no sound or motion to disturb you! Even at that altitude, the controlled atmosphere within the cabin is no thinner than the air at a place such as Colorado Springs.

The gigantic Jet Clippers, costing nearly $6,000,000 each, contain a beautiful, luxurious lounge such as the one pictured, where passengers enjoy silent jet flight, comfort, security, and numerous personal services and conveniences. The fantastic speed of these Jet Clippers makes possible round-trips to Europe in a single day, as well as lower fares, thereby permitting many people who could not ordinarily afford the time or the cost an opportunity to visit other countries. Jet Clippers thus promote the cause of international peace by broadening the understanding of people of all nations as they "sky-hop" from country to country.

The flight deck of a Jet Clipper is actually simpler than the cockpit of a conventional airliner. Jet engines need no propeller controls and therefore many of the usual instruments are unnecessary.

BOEING 727 • A growing need for a jet to fill the gap in air transportation (not covered by the larger jets) to smaller cities located off the main traffic routes, as well as to provide economical through-stop service along main routes, brought about the development of the Boeing 727. Its range of 2,500 miles allows it to make many short flights without refueling. Three turbofan (fanjet) engines, each having a thrust of 14,000 pounds, are

mounted at the rear of the airplane. At an altitude of 37,000 feet, the 727's cruising speed is more than 600 miles per hour. It carries from 71 to 131 passengers. The fuselage is 137 feet 8 inches long; the tail is 34 feet high; the wing span is 108 feet. In the picture above you see a British West Indian Airways 727, which connects various islands in the Caribbean with the mainland of the United States and South America.

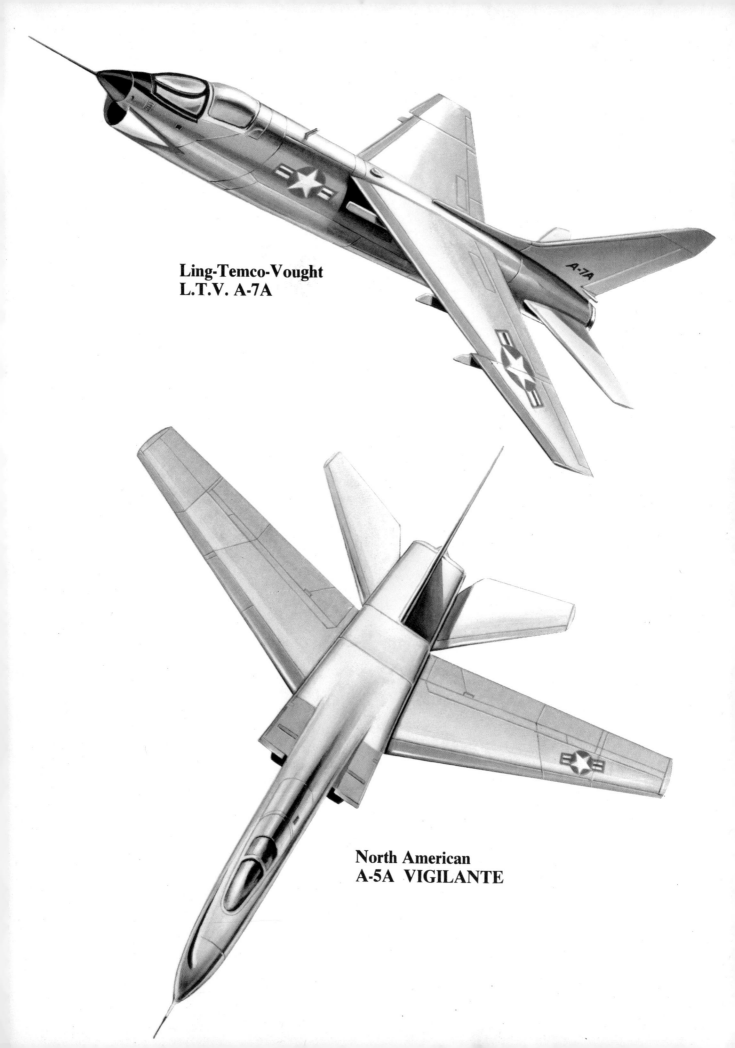

**Ling-Temco-Vought
L.T.V. A-7A**

**North American
A-5A VIGILANTE**

WARPLANES

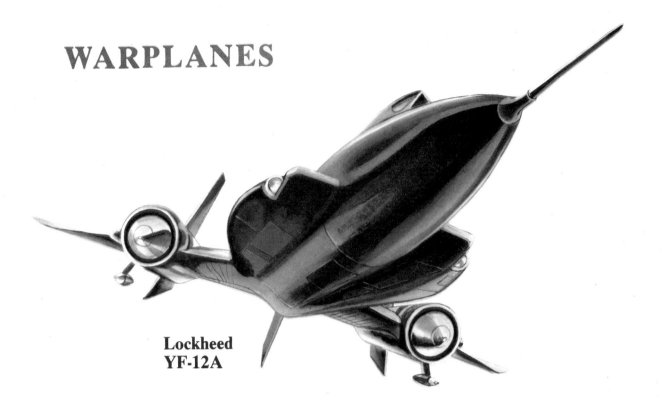

**Lockheed
YF-12A**

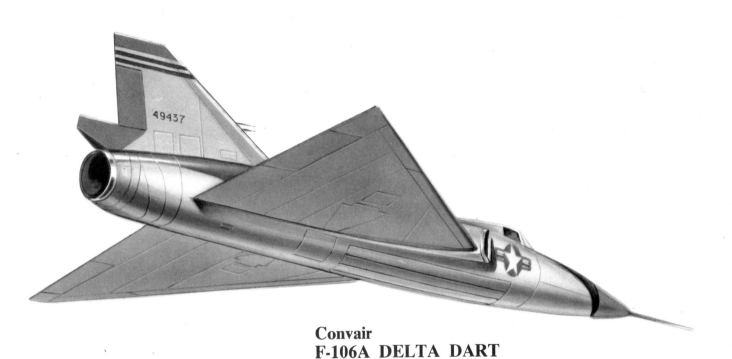

**Convair
F-106A DELTA DART**

"FLYING CRANE" HELICOPTER • The Sikorsky "Flying Crane" can carry a mobile-hospital pod — containing emergency medical equipment and personnel needed to care for soldiers wounded in combat — into the thick of battle. In the picture, it is descending to the side of a mountain on an urgent medical mission. Because of its great lifting ability, this helicopter has virtually opened up a new area of work for itself. It could carry sections of a bridge or a steel tower — or even tow a vessel in distress.

MANNED SPACE ROCKET X-15 • The performance of the X-15 keeps changing as man learns more about space flight. At the present time it is capable of reaching an altitude of 100,000 feet and flying eight times faster than the speed of sound. Under such conditions the outside surface of the craft is heated to a temperature of over 2,400° F. In practice, a B-52 plane (known here as the mother-ship) takes the X-15 to an altitude of about 45,000 feet. The rocket engines are then started, whereupon the X-15 separates from the mother-ship and flies off into space. The yellow tank pod on its side contains extra fuel for longer flights.

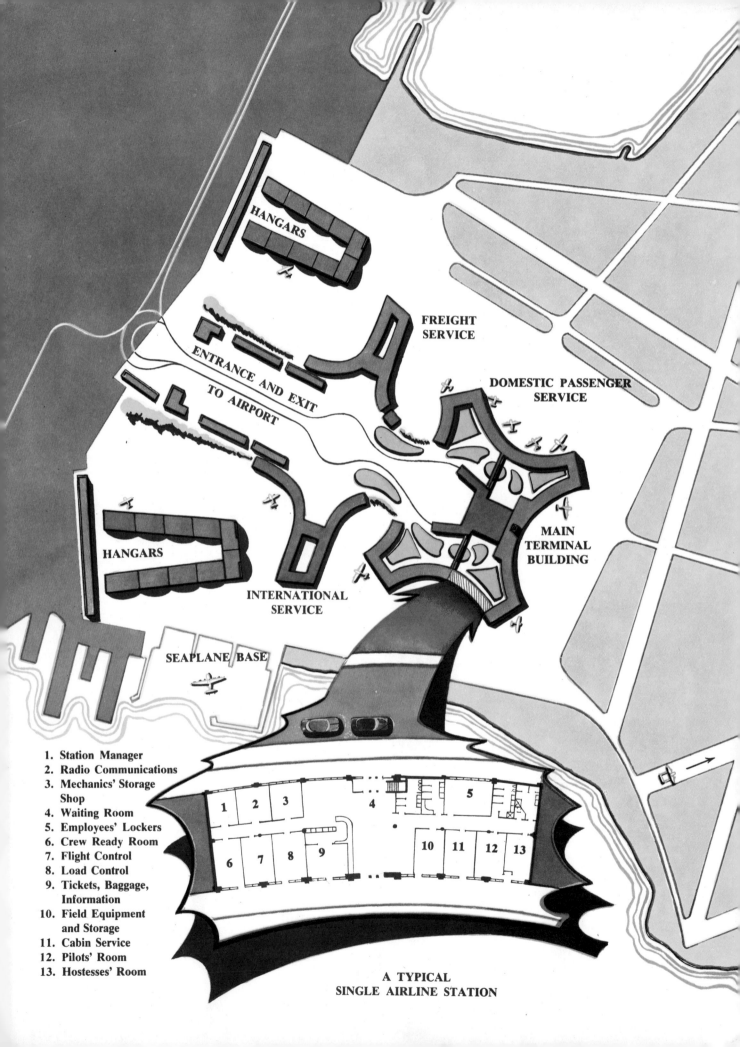

HANGARS

FREIGHT
SERVICE

ENTRANCE AND EXIT
TO AIRPORT

DOMESTIC PASSENGER
SERVICE

HANGARS

MAIN
TERMINAL
BUILDING

INTERNATIONAL
SERVICE

SEAPLANE BASE

1. Station Manager
2. Radio Communications
3. Mechanics' Storage Shop
4. Waiting Room
5. Employees' Lockers
6. Crew Ready Room
7. Flight Control
8. Load Control
9. Tickets, Baggage, Information
10. Field Equipment and Storage
11. Cabin Service
12. Pilots' Room
13. Hostesses' Room

A TYPICAL
SINGLE AIRLINE STATION

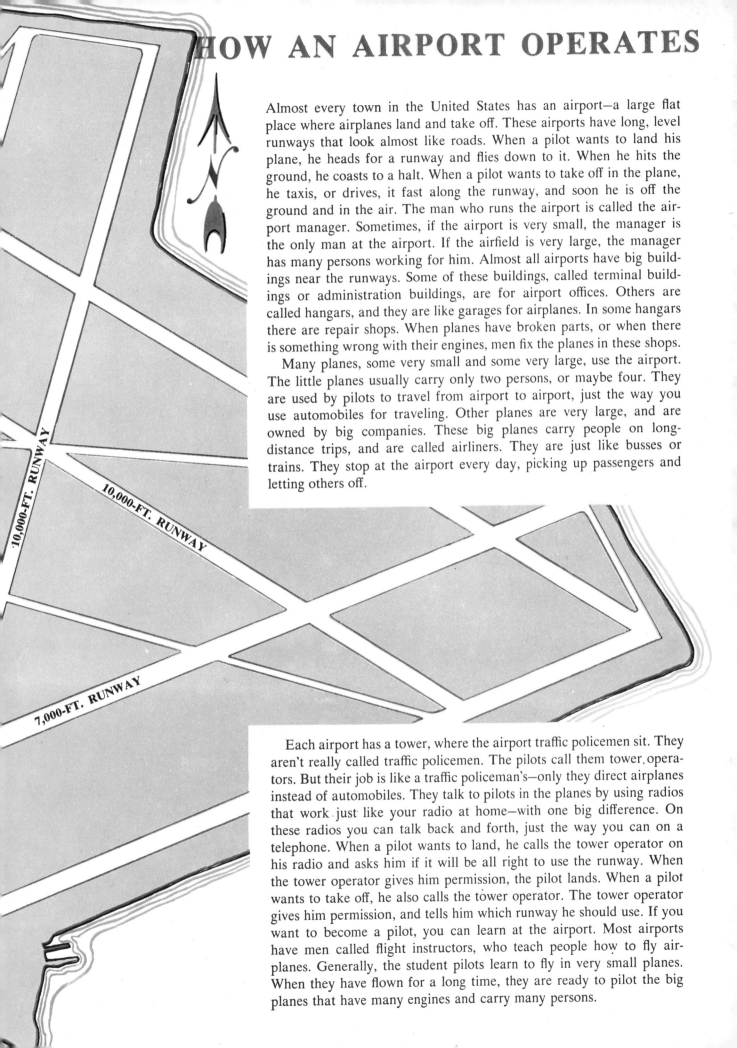

HOW AN AIRPORT OPERATES

Almost every town in the United States has an airport—a large flat place where airplanes land and take off. These airports have long, level runways that look almost like roads. When a pilot wants to land his plane, he heads for a runway and flies down to it. When he hits the ground, he coasts to a halt. When a pilot wants to take off in the plane, he taxis, or drives, it fast along the runway, and soon he is off the ground and in the air. The man who runs the airport is called the airport manager. Sometimes, if the airport is very small, the manager is the only man at the airport. If the airfield is very large, the manager has many persons working for him. Almost all airports have big buildings near the runways. Some of these buildings, called terminal buildings or administration buildings, are for airport offices. Others are called hangars, and they are like garages for airplanes. In some hangars there are repair shops. When planes have broken parts, or when there is something wrong with their engines, men fix the planes in these shops.

Many planes, some very small and some very large, use the airport. The little planes usually carry only two persons, or maybe four. They are used by pilots to travel from airport to airport, just the way you use automobiles for traveling. Other planes are very large, and are owned by big companies. These big planes carry people on long-distance trips, and are called airliners. They are just like busses or trains. They stop at the airport every day, picking up passengers and letting others off.

10,000-FT. RUNWAY

10,000-FT. RUNWAY

7,000-FT. RUNWAY

Each airport has a tower, where the airport traffic policemen sit. They aren't really called traffic policemen. The pilots call them tower operators. But their job is like a traffic policeman's—only they direct airplanes instead of automobiles. They talk to pilots in the planes by using radios that work just like your radio at home—with one big difference. On these radios you can talk back and forth, just the way you can on a telephone. When a pilot wants to land, he calls the tower operator on his radio and asks him if it will be all right to use the runway. When the tower operator gives him permission, the pilot lands. When a pilot wants to take off, he also calls the tower operator. The tower operator gives him permission, and tells him which runway he should use. If you want to become a pilot, you can learn at the airport. Most airports have men called flight instructors, who teach people how to fly airplanes. Generally, the student pilots learn to fly in very small planes. When they have flown for a long time, they are ready to pilot the big planes that have many engines and carry many persons.

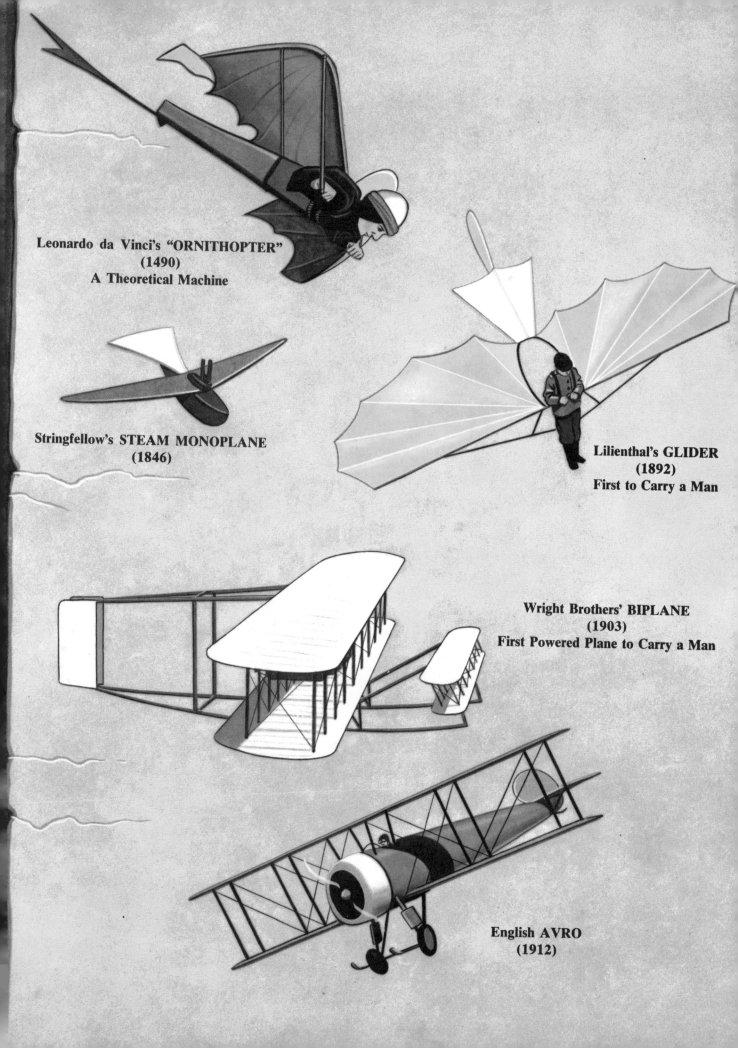

Leonardo da Vinci's "ORNITHOPTER"
(1490)
A Theoretical Machine

Stringfellow's STEAM MONOPLANE
(1846)

Lilienthal's GLIDER
(1892)
First to Carry a Man

Wright Brothers' BIPLANE
(1903)
First Powered Plane to Carry a Man

English AVRO
(1912)